Let's Look at Lizards

IGUANAS

By Seth Lynch

Gareth Stevens
PUBLISHING

Please visit our website, www.garethstevens.com. For a free color catalog of all our high-quality books, call toll free 1-800-542-2595 or fax 1-877-542-2596.

Library of Congress Cataloging-in-Publication Data
Names: Lynch, Seth, author.
Title: Iguanas / Seth Lynch.
Description: Buffalo, New York : Gareth Stevens Publishing, [2025] | Series: Let's look at lizards! | Includes bibliographical references and index.
Identifiers: LCCN 2023034804 | ISBN 9781538292761 (library binding) | ISBN 9781538292754 (paperback) | ISBN 9781538292778 (ebook)
Subjects: LCSH: Iguanas–Juvenile literature.
Classification: LCC QL666.L25 L96 2025 | DDC 597.95/42–dc23/eng/20230803
LC record available at https://lccn.loc.gov/2023034804

First Edition

Published in 2025 by
Gareth Stevens Publishing
2544 Clinton Street
Buffalo, NY 14224

Editor: Brianna Propis
Designer: Leslie Taylor

Photo credits: Cover Kurit afshen/Shutterstock.com; p. 5 Ridho Arifuddin/Shutterstock.com; p. 7 Chaikom/Shutterstock.com; p. 9 (top) Maria Maarbes/Shutterstock.com, (lower left) Kurit afshen/Shutterstock.com, (lower right) EVA SUKMA H/Shutterstock.com; p. 11 dwi putra stock/Shutterstock.com; p. 13 buteo/Shutterstock.com; p. 15 Ronsmith/Shutterstock.com; p. 17 Jeff Grabert/Shutterstock.com; p. 19 GUDKOV ANDREY/Shutterstock.com; p. 21 J.NATAYO/Shutterstock.com; p. 23 Halfpoint/Shutterstock.com.

Printed in the United States of America

CPSIA compliance information: Batch #CSGS25: For further information contact Gareth Stevens, New York, New York at 1-800-542-2595.

Contents

Which lizard is this?
It's an iguana!

They have scales.
These are bumpy.

They can be green or blue.
They can be orange
or black.
They change color
in the sun.

They have a long tail.
It can fall off.
It can grow back!

They live in trees near water.
They like to swim!

They are good jumpers.
They jump into water!

They eat plants and fruit.
They eat leaves
and flowers.

They have sharp teeth.

Mothers lay eggs.
They keep the eggs safe.

They can live in the wild. They can be kept as a pet.

Words to Know

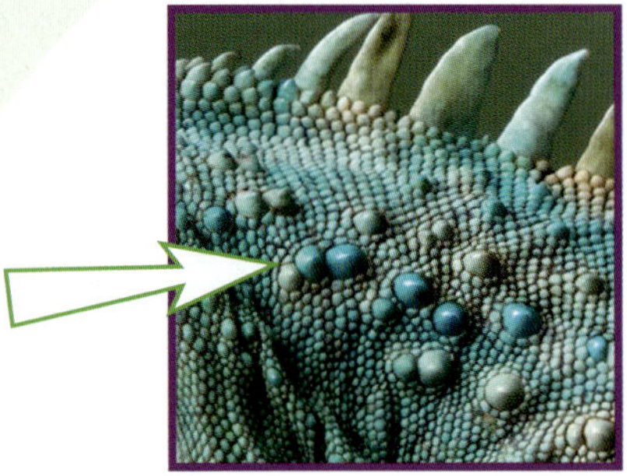

bumpy

egg

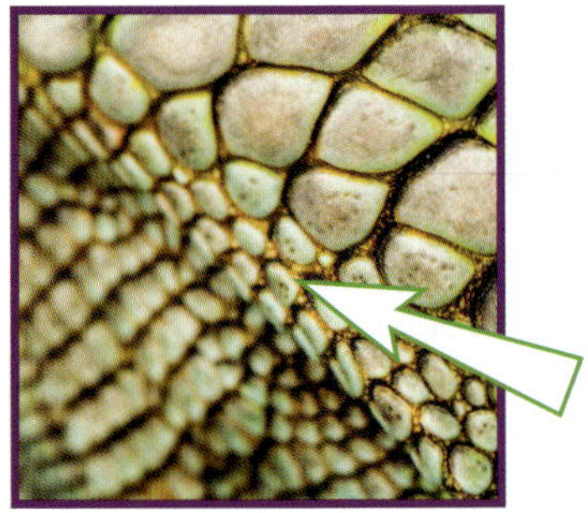

scales

Index